More Science Surprises

from
Dr. Zed

Gordon Penrose
Edited by Marilyn Baillie

Simon & Schuster Books for Young Readers
Published by Simon & Schuster
New York   London   Toronto   Sydney   Tokyo   Singapore

SIMON & SCHUSTER BOOKS FOR YOUNG READERS, Simon & Schuster Building
Rockefeller Center, 1230 Avenue of the Americas, New York, New York 10020
Copyright © 1992 by Greey de Pencier Books. All rights reserved including the right of
reproduction in whole or in part in any form. Originally published in Canada by Greey
de Pencier Books. First U.S. edition 1992. SIMON & SCHUSTER BOOKS FOR YOUNG READERS is
a trademark of Simon & Schuster. Photography by Ray Boudreau, Illustrations by
Tina Holdcroft. Design concept by Wycliffe Smith. Designed by Julia Naimska.
Manufactured in Hong Kong.

10 9 8 7 6 5 4 3 2 1

(pbk) 10 9 8 7 6 5 4 3 2 1

Library of Congress Cataloging-in-Publication Data, Penrose, Gordon. More science
surprises from Dr. Zed / Gordon Penrose: edited by Marilyn Baillie.   p.   cm.
Includes index. Summary: An illustrated collection of simple and safe science
experiments.   1. Scientific recreations—Juvenile literature.   2. Science—
Experiments—Juvenile literature. [1. Science—Experiments.   2. Experiments.
3. Scientific recreations.] I. Baillie, Marilyn.   II. Title.   Q164. P363
1992 507.8—dc20 91-38935 CIP

ISBN 0-671-77810-2
ISBN 0-671-77811-0 (pbk)

Some of the material in this book was previously published in Chickadee Magazine.

The activities in this book have been tested and are safe when conducted as
instructed. The publisher accepts no responsibility for any damage caused or
sustained due to the use or misuse of ideas or materials featured in this book.

Contents

Hot and Cold

**Can you make a balloon blow up
without huffing and puffing?**

HERE'S HOW:

1. Remove the cap from an empty glass or plastic soda bottle and put the bottle in the freezer for 15 minutes.

2. Blow up a small balloon to stretch it. Let the air out again.

3. Take the bottle out of the freezer, and pull the end of the balloon over the bottle opening.

4. Hold the bottle under your arm and watch the balloon blow up!

HAVE YOU EVER HEARD A BOTTLE TAPPING? FOLLOW THESE STEPS AND THEN LISTEN...

Put an empty soda bottle in the freezer for half an hour. Then take the bottle out and run water over the opening. Place a wet coin on top of the opening and cup your warm hands around the bottle. Now listen for the tap, tap, tap.

Water Surprises

Play the master magician and perform the magic arrow flip trick.

HERE'S HOW:

1. Draw a big arrow on an index card.

2. Hold the card upright a hand's length behind an empty glass. Look at the arrow.

3. Fill the glass with water. PRESTO! The arrow points the other way!

> It must be magic!

THE DISAPPEARING STRAW

Fill a plastic glass with water. Put a straw in the water and look at it through the side of the glass. Move the straw slowly around the inside of the glass. Wow! Two straws! Move it some more. Now no bottom straw. Keep moving the straw around to find it again...all in one piece.

Up and Away!

It's a bird! It's a plane! No, it's a soaring straw!

1. Tape a paper clip to one end of a straw.

2. Cut two strips from the short side of a piece of 8½'' x 11'' paper.

3. Tape the strips into loops. Then tape one loop to each end of the straw. (See picture.)

4. Hold the plane with the paper clip end forward and the loops up, and launch it. The plane flies!

Make Zoomers with your friends and see whose sails the farthest!

ZOOMERS

Cut a big straw in half. Tape one end of one piece closed. Now slide this half over the end of a thinner straw. Blow through the thin straw. Zoom! The big straw is off and away.

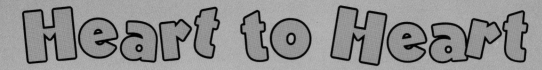

Heart to Heart

**Roses are red, violets are blue,
Hugs and twirls, from me to you.**

Hugging Hearts

1. Trace the heart pattern at left onto a piece of paper.

2. Cut out the traced pattern, fold it over the middle of a pencil and tape it in place.

3. Can you make the hearts hug by blowing in between them?

WHIRLING TWIRLER

Fold an index card in half. Draw a large red heart on one side and a smiling face on the other. Now put the folded card on the end of a pencil. To hold the card in place, staple it together on both sides of the pencil. Gently hold the pencil between your palms and rub them together.

Give it a whirl!

Egg-citing Eggs

Make eggs bright and beautiful by adding a special ingredient.

HERE'S HOW:

1. Fill a small bowl with water and add ten drops of food coloring.

2. Place a hard-boiled egg in the water for about two minutes. Take the egg out and look at its color.

3. Now add three large spoonfuls of vinegar to the bowl.

4. Put the egg back in the bowl for two more minutes. Look at its color now. Wow! What a change.

The Stand-Up Egg

1. Place a large spoonful of salt on a counter near a sink.

2. Stand an egg up in the salt.

3. Gently blow the salt away toward the sink. Amazing! The egg stays standing.

TAKE A TWIRL TEST

Can you tell if an egg is hard-boiled or raw without cracking its shell? Just spin the egg on its side. A hard-boiled egg will spin much faster than a raw one.

13

Bubble, Bubble

Club soda can make your food do some funny things!

Dancing Raisins

1. Fill a glass with club soda.
2. Drop a few raisins into the glass. Wow! Watch them dance.

Swimming Spaghetti

1. Fill a glass with club soda and add a little blue food coloring.
2. Now drop in some pieces of dry uncooked spaghetti. Up they swim and down they dive into the deep blue sea.

DO YOU KNOW WHAT RAISINS USED TO BE?

Leave your raisins in club soda overnight. Take a look at them the next morning and see what you find.

Paper Power

Which of these paper shapes will hold up a book? Do some folding to find out.

1 Fold or roll pieces of paper into the shapes in the picture.

2 Try to balance a book on each shape. Now make up other shapes and try them, too.

Ready, set, go! Which piece of paper will win the Great Paper Race?

1 Find two sheets of paper that are exactly the same size.

2 Crumple one into a ball and leave the other flat.

3 Stand on a chair and drop the two pieces at the same time. Which one gets to the floor first? Race them again and see which one wins this time.

One, two, three... PULL!

PAPER CHALLENGE: TRY TO TEAR THIS PAPER INTO THREE.

Make two identical slits on the top of a piece of paper. Now hold the paper on each side and pull out to tear it. What happens?

Finger Fun

The Incredible Dry Finger

1. Fill a glass halfway with water and sprinkle a thick layer of baby powder on top.

2. Slowly stick your finger down into the water.

3. Lift your finger straight up out of the glass. Incredible! It's dry!

The Runaway Pepper

1. Fill a glass with water and sprinkle some pepper on it.

2. Dip your finger in liquid detergent and stick your finger in the middle of the glass. Wow! The pepper shoots away.

Make your own "band" music!

FINGER MUSIC

Put a rubber band around a drinking glass from top to bottom. Now place the open end of the glass close to your ear. Pluck the band with your finger and listen.

Bathtub Boats

Build a boat that will travel your tub.

1. Ask an adult to cut a one-pint milk carton in half lengthwise.

2. Staple a plastic straw on either side of the carton as shown in the diagram.

3. From the leftover carton, cut out a rectangle about two thumbs high and one thumb wide.

4. Slip a rubber band around the rectangle and staple securely in the middle on both sides.

5. Slip the rectangle's rubber band over the ends of the two straws. This makes a propeller.

6. Bend both ends of a third straw and slide them into the ends of the two side straws to form a frame for the propeller.

Launch Your Boat

Put your boat in water, wind up the rubber band, and away it goes!

milk carton

Staple straws.

propeller
↓

Staple rubber band.

rubber band

Slide straws together.

Look at it go!

THE SPEEDBOAT

Cut a simple boat shape out of aluminum foil. Now dab a little liquid detergent on the back edge of the boat. Put the boat on soap-free water and step back for the takeoff!

Bag Boggle

Wow! It's the no-leak lunch bag trick!

1. Fill a plastic sandwich bag with water and knot it.

2. Hold the bag over a sink with one hand.

3. Push a pencil through one side of the bag and out the other. Ta-da! No leaks.

See how many more pencils you can push through.

NO-BUDGE LUNCH BAG

Open up a plastic sandwich bag inside a glass and push it tightly against the bottom and sides. Next, bring the top of the bag up and over the rim of the glass and secure it with a rubber band. Reach to the bottom and try to pull the bag out of the glass. It's there to stay!

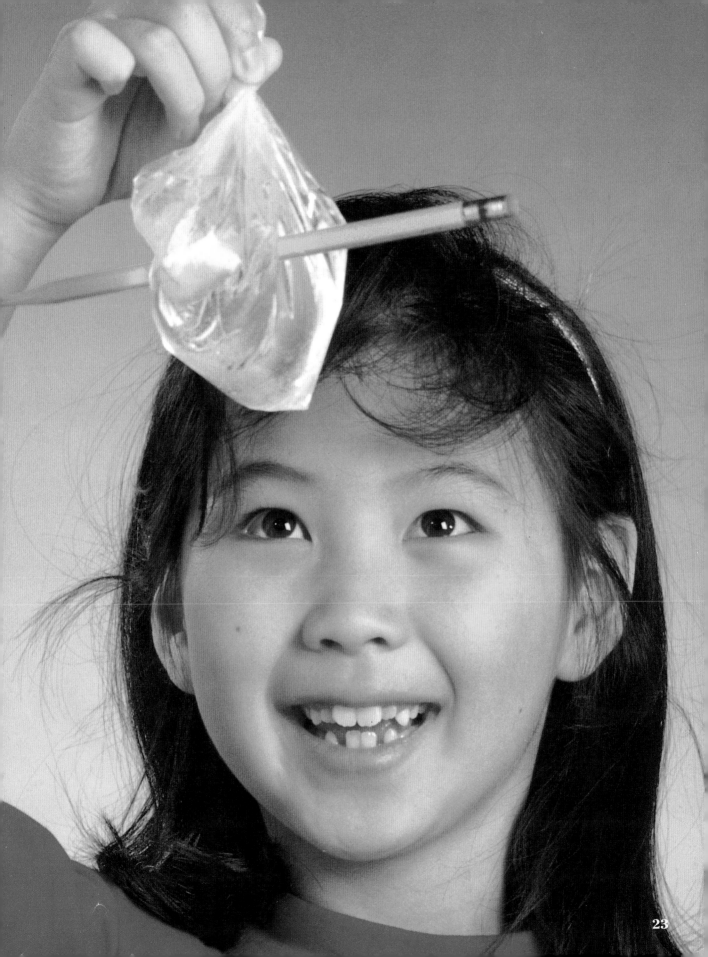

Plant Power

Grow some light-seeking, hole-peeking beans.

HERE'S HOW:

1) Cut a circle the size of a small juice can in one end of a shoe box.

2) Plant two dried navy or kidney beans in a small plastic container filled with soil.

3) Stand the container in the end of the box opposite the hole.

4) Put the lid on the box and set it in a sunny place. Check it every day and water the beans every few days.

5) When your plant begins to peek through the hole, take off the lid to see how it grew. We've snipped away the side of the box to show you what happened.

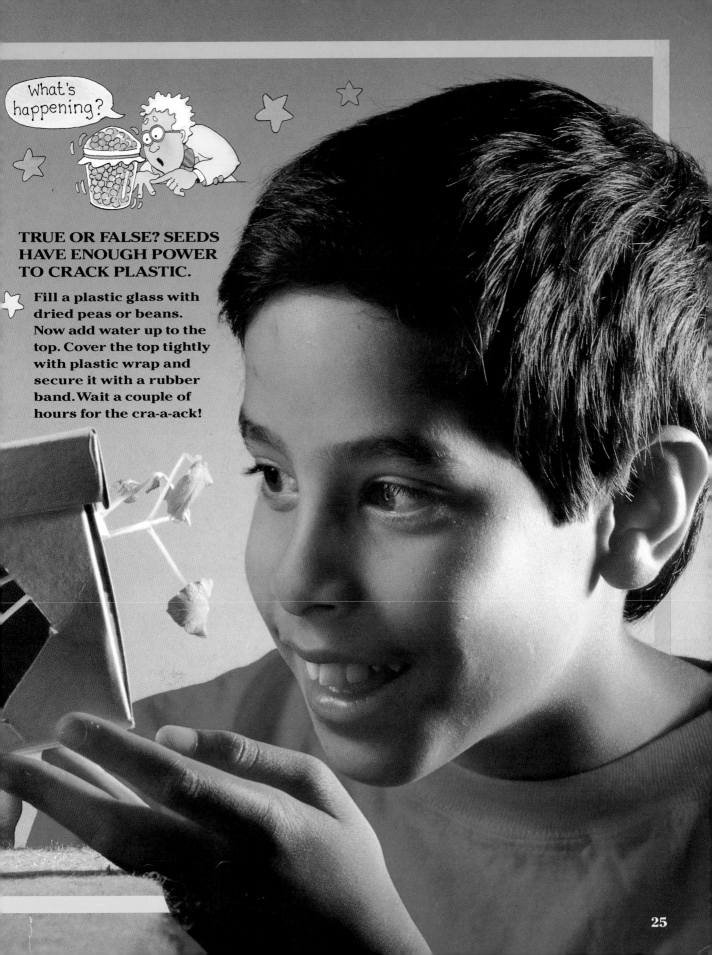

What's happening?

TRUE OR FALSE? SEEDS HAVE ENOUGH POWER TO CRACK PLASTIC.

Fill a plastic glass with dried peas or beans. Now add water up to the top. Cover the top tightly with plastic wrap and secure it with a rubber band. Wait a couple of hours for the cra-a-ack!

Night Lights

Have you ever wondered if the moon changes in size? Follow these steps to see if it really does.

1. Watch for a full moon. When it first rises, the moon often looks very large. Look at it through a hole in three-ring-binder paper.

2. Later, when the moon is high in the sky and appears to be smaller, look at it again through the same hole. It's the same as before!

STAR LIGHT, STAR BRIGHT

Take a good look at each of these stars. Which one looks bigger to you?

Try this eye test!

Color Vision

How much do you know about colors?
Is black ink really black?

1. Draw a heavy black dot with a water-soluble marker on a paper towel or coffee filter.

2. Now put a drop of water directly on top of the dot.

3. Watch the dot come apart. What colors do you see?

Three colors...or more?

1. Mix a drop each of red, yellow, and blue food coloring in a small amount of water.

2. Fold a coffee filter in half and then into quarters. Place the narrow end in the mixture.

3. Let the filter soak up the mixture, then open it up. So many colors!

Try bubbling a second color over the first.

SPATTER PATTERNS

Mix a small spoonful of detergent, a large spoonful of water and eight drops of food coloring in a small glass. Now hold the glass over some paper and blow into the mixture with a straw. Move the glass around as the bubbles spill onto the paper. What spatter patterns!

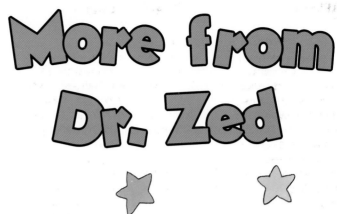

More from Dr. Zed

There's SCIENCE everywhere!

The activities in this book have been carefully selected to give children a hands-on introduction to science. In doing these simple experiments, children can make a variety of discoveries that will surprise and delight them. Above all, it is hoped that these activities will encourage children to explore and experiment further, and to discover for themselves the many surprises science holds.

There are no wrong results to the activities—every finding is valid, and bound to stimulate lots of questions. The following notes suggest explanations and insights into those science concepts that can be discussed simply with young children; questions about more complex concepts are better explored through hands-on experimentation than by explanations too abstract for the very young. These notes also include the answers to special challenges scattered throughout the book.

Hot and Cold
(page 4)
Activities #1 and #2

Cold air takes up less room than hot air. When you heat up the cold air in the bottle with your body heat, the warmed air expands and takes up more space. In Activity #1 this escaping air has nowhere to go but into the balloon. The warmed air in Activity #2 escapes out the top a little at a time, lifting the coin up and dropping it.

Water Surprises
(page 6)
Activities #1 and #2

Water is clear, but when you look through it, it acts as a lens, and changes the way you see things. In Activity #1 looking through the water flips the image so that it looks backward to you. In Activity #2 the straw appears to separate or disappear as you move it around because you are looking across at the straw through the water. Pour the water out and the straw stays straight and in one piece.

Heart to Heart
(page 10)
Activity #1

When you blow between the hearts, you create a space where there isn't much air. The air outside the hearts rushes into that empty space and pushes the hearts in with it, making them touch.

Activity #2

The card spins so fast that your eye still holds a picture of the first side when the second side appears. You end up seeing both sides together. This is called imaging.

Egg-citing Eggs
(page 12)
Activity #1

An eggshell is smooth, so food coloring can't stick to the shell very well. Vinegar is an acid, which eats away at the outside of the shell, leaving it rough and porous. Now more of the food coloring can soak into the shell.

Activity #2

When you stand the egg up, it is supported by just a few grains of salt that are wedged underneath it. The rest of the salt can be blown away because only these few grains are doing the job.

Activity #3

When you spin a raw egg, the liquid inside it moves around, slowing the egg down. The hard-boiled egg spins faster and better because it is solid inside.

Bubble, Bubble
(page 14)
Activities #1 and #2

Some of the bubbles in the club soda collect on the raisins. The bubbles rise to the top carrying the raisins with them. When the bubbles reach the top and burst, the raisins sink again. The bubbles also lift the spaghetti pieces, but because they are lighter they rise more easily.

Activity #3

Raisins were once fat juicy grapes that were picked and dried in the sun. When your raisins fill up with water or club soda, they take on their former grape shape.

Paper Power
(page 16)
Activity #1

The accordion shape and the tube hold up the book the best. The open book shape and the tent shape collapse under the heavy book because one part of the paper shape is weaker and gives way. Which shape is the best? Press down and you'll find that the tube is the strongest. This shape is often used in construction because it is so strong.

Activity #2

Air presses in all directions. The flat piece of paper has much more surface exposed for the air to resist. This slows the flat piece down and the crumpled piece wins every time.

Activity #3

Why does the paper always rip into only two pieces? As you pull the paper sideways, your pull is never exactly the same on both sides. One is weaker and one is stronger. And your two cuts are never exactly the same although they may look identical. For these two reasons, your paper will tear at either one cut or the other. As you pull, it continues to tear down the weaker side.

Finger Fun
(page 18)
Activity #1

Your finger is a little bit oily from natural oil in your skin, so the powder floating on the water sticks to the oil on your finger. As you push your finger down into the water, the powder acts like a glove. It repels the water and keeps your finger dry.

Activity #2

Water behaves as if it has a clear elastic skin on it. The soap breaks the skin, which pulls the pepper grains to the sides of the glass.

Activity #3

Sounds are rapid vibrations in the air. When you pluck the rubber band, you start the air moving in the tumbler. The air passes the vibrations to your ear, which acts like a funnel to collect the sound.

Bathtub Boats
(page 20)
Activity #1

When you wind up the rubber band, the energy you use to wind is stored in the band. When you let go, the band unwinds, letting out that stored (potential) energy. This turns the propeller and pushes the boat through the water.

Activity #2

The water behaves as if it has a clear elastic skin on the surface. The soap from the boat breaks the skin and the skin pulls back toward the sides of the basin, taking the boat with it.

Bag Boggle
(page 22)
Activity #1

When you pierce the plastic bag, the plastic pulls together around the pencil to seal up the hole. Then no water can escape.

Activity #2

When you press the plastic bag against the inside of the glass, there is less air between the bag and the glass than there is pressing in from above and outside the glass, so you can barely budge the bag. If you put a tiny hole in it, air rushes in underneath it and out comes the bag.

Plant Power
(page 24)
Activity #1

Beans, like other green plants, need water and light to make food and grow. The only light is through the hole at the end, so the beans grow in that direction.

Activity #2

Seeds take in water, swell, and break open as they grow. As your beans get bigger, they need more room to expand, and crack the plastic glass.

Night Lights
(page 26)
Activity #1

The moon fits in the hole both times. Some scientists think that the moon seems to change size because you can compare it with trees and buildings when it has just risen. When it is high in the sky it just seems smaller because there are no large forms to compare it with.

Activity #2

The stars are both the same size. This is an optical illusion, which is a picture that tricks your eyes and fools your brain.

Color Vision
(page 28)
Activities #1 and #2

As the drop of water spreads out, it takes some of the ink with it. Black ink is made up of many colors. The different colors separate out of the spot of ink. The "heavy" colors stay near the center. "Light" ones spread out to the outside. In the same way, your murky mixture separates out in the coffee filter.